Lynx

The Final Year in French Service

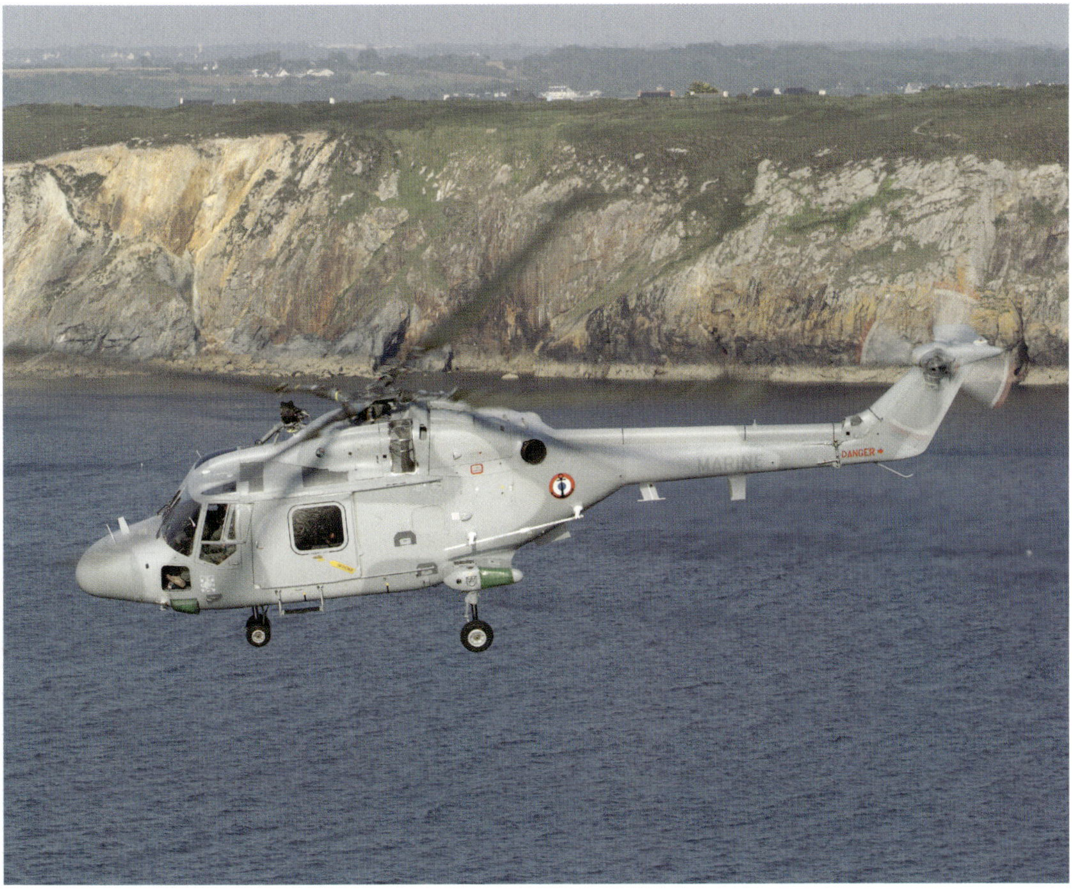

HENRI-PIERRE GROLLEAU

KEY Books

MODERN MILITARY AIRCRAFT SERIES, VOLUME 1

Front cover image: A Flottille 34F Lynx hovers above the Atlantic Ocean, off Brittany. The final year of operations was intense for the squadron.

Back cover image: Both the UK and France have now withdrawn all their Lynxes. However, the type will soldier on with many other nations for years to come.

Title page image: The Lynx was a common sight off Brittany for 40 years. It will be missed by many.

Contents page image: French Navy Lynxes enjoyed an extremely successful career. Although a pure product of the Cold War, they were never engaged in high-end combat ops.

Published by Key Books
An imprint of Key Publishing Ltd
PO Box 100,
Stamford
Lincs, PE19 1XQ

www.keypublishing.com

Acknowledgements

The author would like to thank all military and civilian personnel for the invaluable assistance provided during his various reports, at Lanvéoc-Poulmic and on board frigate *Latouche-Tréville*: Admiral Goutay, Admiral Janicot, Captain Berling and all BAN Lanvéoc-Poulmic personnel, Commander Chaput, Lieutenant Commander Sébastien, Lieutenant Commander Christophe, Lieutenant Jean-André and all Flottille 34F personnel, Captain de Sevin and all *Latouche-Tréville* personnel, Captain Lavault, Lieutenant Commander Olivier, Lieutenant Emilie, Sub-Lieutenant Vincent and Mrs Corre.

All photographs are the copyright of Henri-Pierre Grolleau.

The right of Henri-Pierre Grolleau to be identified as the author of this book has been asserted in accordance with the Copyright, Designs and Patents Act 1988 Sections 77 and 78.

Copyright © Henri-Pierre Grolleau, 2020

ISBN 978 1 913870 12 6

All rights reserved. Reproduction in whole or in part in any form whatsoever or by any means is strictly prohibited without the prior permission of the Publisher.

Typeset by SJmagic DESIGN SERVICES, India.

Contents

Foreword		4
Chapter 1	The Lynx in French Navy Service	5
Chapter 2	Lynx Testers	24
Chapter 3	Flottille 34F and the Lynx	26
Chapter 4	French ASW Frigates	43
Chapter 5	Special Forces Lynx	59
Chapter 6	Lynx SAR Missions	65
Chapter 7	Maintaining a Proven Rotorcraft	74
Chapter 8	Final Lynx Deployment at Sea	82

Foreword

Ten years after the retirement of the mighty SA321G Super Frelon, the iconic Lynx has now also been withdrawn from Aéronautique Navale service. The Westland Lynx was the end result of a comprehensive and successful Franco-British cooperation that also produced the Gazelle, the Puma and the Jaguar. When introduced in 1978, it constituted a massive improvement over the older Sikorsky HSS-1 and Aérospatiale Alouette III helicopters then in service with the French Navy for anti-submarine duties. It was fitted with a modern weapon system that comprised a very effective French-sourced sonar and was specifically conceived by British designers to operate in all weathers, day and night, from pitching frigate and destroyer decks.

Within the French Navy, the Lynx was a highly popular rotorcraft. It was considered to be a robust, reliable, dependable and highly effective helicopter. Even more importantly, in difficult weather conditions, when winching survivors or fast roping commandos onto a ship, it was powerful, responsive and remarkably agile. Pilots just loved it! From an asset specialising in anti-submarine warfare, it was soon turned into a multirole platform that proved incredibly versatile, performing an extremely wide range of roles, at sea and from its various home bases. But the Lynx was an ageing design, and its successor, the NH90 Caïman, was already there to take over its various missions. Farewell Lynx! You will be sorely missed!

Henri-Pierre Grolleau

Lynx s/n 622 in the hover near the Pierres Noires (black stones) lighthouse, off the western tip of Brittany, in June 2020.

Chapter 1
The Lynx in French Navy Service

On 4 September 2020, the iconic Lynx was officially withdrawn from use by the French Navy after 42 years of sterling service, thus leaving the NH90 Caïman as the only Marine Nationale anti-submarine helicopter.

The advent of the Lynx, in the late 1970s, had allowed the French Navy to refine its new anti-submarine warfare (ASW) concept that relied on dedicated ASW frigates and helicopters to detect, track and attack hostile submarines. Capitaine de Frégate (Commander) François Chaput, Commanding Officer of Flottille 34F, the last operational Lynx naval squadron in France, explains:

> When the Force Océanique Stratégique, the submarine nuclear deterrence force, was set up in the late 60s, ASW became a high-priority mission to escort and protect *Redoutable*-class SSBNs. Sikorsky HSS-1 helicopters were modified for the role and aircrews developed initial ASW tactics using rotorcraft. With the advent of the Super Frelon and of the Alouette III, tactics and

The Alouette III was used in anti-submarine warfare by the Marine Nationale in the 1970s. Today, a few Alouettes soldier on in the light support role with Escadrille 22S, a unit that is due to be renumbered Flottille 34F in early 2021.

operational procedures were progressively refined. But these aircraft had obvious shortcomings: from 1974, the single-engine Alouette III used a MAD, a magnetic anomaly detector, to try to detect a submarine, a system inherently less efficient than a dunking sonar. If a submarine had been detected, it could have been attacked with either a US-made Mk 44 torpedo, or depth charges. The three-engine Super Frelon was more effective as it offered a longer range and a much better payload. It could fly in all weathers, day and night, but it was big and heavy and could not operate from the pitching deck of a frigate. As a result, it could only be flown from the *Clemenceau* and *Foch* aircraft carriers, and from the *Jeanne d'Arc* helicopter-carrier. These three types gave sailors the opportunity to develop an operational doctrine and to progressively improve their tactics. The entry into service of the Lynx equipped with a DUAV-4 sonar and with Mk 46 torpedoes represented a decisive step forward.

Two Variants into French Navy Service

Two French development aircraft, XX904/F-ZKCU and XX911/F-ZKCV, took part in the multinational fly test programme. Two production variants of the Lynx were accepted into Marine Nationale service:

- 26 Lynx Mk 2s (s/n 260 to 267, 269 to 278 and 620 to 627) powered by Gem 2 turboshafts;
- 14 Lynx Mk 4s (s/n 801 to 814) powered by more powerful Gem 41-1 engines.

The French Mk 2 was broadly similar to the early Royal Navy Lynx HAS2 but with different avionics and mission suite and a higher all-up weight (10,500lbs/4,763kg instead of 9,750lbs/4,423kg). While the HAS2 specialised in anti-ship warfare with its Seaspray radar and its Sea Skua missiles, the French Mk 2 was optimised for the ASW mission and was equipped with an ORB-31 radar, a DUAV-4 dunking sonar and Mk 46 torpedoes. It could also undertake limited anti-surface operations with the optically guided AS12 short-range air-to-surface missile. Thanks to its new engines, the Mk 4 offered expanded capabilities and a maximum take-off weight boosted to 10,750lbs (4,876kg). All surviving Mk 2s were eventually upgraded to Mk 4 standard.

At its peak, three front-line units operated the type: Flottille 31F, at Saint-Mandrier, and Flottilles 34F and 35F, both at Lanvéoc-Poulmic. During the Lynx's 42-year career, four helicopters were lost, claiming the lives of 13 personnel.

Unprecedented Capabilities at the Time

Michel Del Guidice was one of the first French Navy pilots to transition to the Lynx. At the time, he had logged 2,200 hours on the Sikorsky S-58/HSS-1 and would eventually log a further 800 hours on the Lynx before leaving the French Navy in 1985. He had accrued considerable ASW experience flying the HSS-1, the first fully effective and reliable helicopter type to enter Aéronavale service. The S-58s were initially built in the USA, but a production line was soon established by Sud-Aviation, at Marignane, in the south of France. For anti-submarine warfare, the HSS-1 was first equipped with an AQS-4 dunking sonar, plus Mk 43 torpedoes and Mk 54 depth charges. From 1969, the anti-submarine system was modernised with the introduction of the much more effective DUAV-1 sonar and of the new Mk 44 torpedo. Michel Del Guidice recalls:

Photographed in 2009, this sonar operator monitors the operation of the sonar reeling system during an ASW training sortie. Noteworthy is the totally outdated plotting board used to manually report targets.

We routinely conducted ASW exercises. Our ASW capabilities were in fact rather limited, with sonar detection ranges varying between 800 and 1,500 yards only. This meant that we needed at least three HSS-1s to track a submarine with their sonars and a fourth one to drop a torpedo. Tactics were extremely complex and sometimes rather ineffective. For us, the arrival of the Lynx was a major step forward in combat capabilities. I was involved in the type's operational evaluation at Fréjus-Saint Raphaël with Escadrille 20S right from the beginning. We had a few teething problems, but service entry was rather smooth. At the time, we were flying the Lynx Mk 2 which was underpowered: an engine failure in the hover was a major emergency as we would have crashed or ditched because the remaining turbine was not powerful enough for us to perform a fly away safely. This issue was later solved with the appearance of the Mk 4. In any case, the type was much safer than the piston-powered HSS-1.

The Lynx was significantly easier to operate than the much older HSS-1. Michel Del Guidice continues:

On the 'Siko', we had a throttle on the collective lever to manually adjust power so that rotor rpm could remain within the required range. Power regulation was entirely automatic on the Lynx and that was a major step forward that helped noticeably reduce aircrew workload. The Lynx was also much faster and substantially more manoeuvrable than the HSS-1.

Operational Revolution

When the first Lynx was delivered by Westland to the French Navy in 1978, it signalled the beginning of an operational and doctrinal revolution. François Chaput explains:

The aircraft impeccably fulfilled the requirement as it was perfectly integrated with our ASW frigates and could operate in sea state 6. It brought to the fleet stand-off ASW capabilities which allowed enemy submarines to be engaged day and night at long distances, before they could threaten the frigate and the ship, or the ships, she was escorting. During the Cold War, Lynx detachments had two aircraft each, giving each frigate a considerable rotary force that could operate round the clock to detect, track and attack hostile submarines. More importantly, F70-class frigates hangars and platforms had been specifically designed to accommodate the Lynx, with a massively positive impact on operational efficiency.

The Lynx progressively became the first truly and fully multirole naval helicopter in service in France. For example, until the advent of the SA365F Dauphin in 1990, the type handled plane guard-duties at night from aircraft carriers *Clemenceau* and *Foch*, ready to winch out the pilot/crew member of any F-8E(FN) Crusader, Étendard IVP, Super Étendard or Alizé that might have ejected or ditched. 'I like to say that the Lynx was the first "total naval helicopter" because it could fly all types of missions, day and night, in a very wide range of weather conditions', insists Commander Chaput. He continues:

> On top of its traditional ASW role, it could be tasked with beyond-the-horizon targeting and target designation for Exocet anti-ship missiles launched by the frigate, maritime counter-terrorism, search and rescue, fishery protection, traffic interdiction, vertical replenishment, medical evacuation, troop transport and embargo control, as in the Adriatic during the war in Bosnia, or in the Eastern Mediterranean, off Syria. With the withdrawal of the Lynx, the French Navy has, in some ways, lost its airborne Swiss Army knife.

State-of-the-Art Torpedo

For the anti-submarine mission, the Lynx initially carried up to two US-made Mk 46 torpedoes. From 2010, they were replaced with the much newer Franco-Italian MU90 Impact torpedo. A number of modifications had to be implemented for the Lynx to become MU90-capable: the electrical system was modified, and the torpedo control unit was altered so that aircrews could programme the weapon before firing. Depending on the sea state, the desired search pattern, the selected initial torpedo depth and the requested firing mode, pre-set data was transferred from the Lynx to the MU90 via the new torpedo control unit. The MU90 could be released either in forward flight or in the hover.

The latest generation of submarines are quieter, faster and capable of diving to greater depths than their predecessors. Recent diesel/electric submarines are often fitted with air-independent propulsion that allows non-nuclear boats to operate without access to oxygen, giving them considerable submerged endurance and an excellent mobility to avoid detection and defeat attacks. The state-of-the-art MU90 gave the Lynx the ability to hit double-hulled, acoustically coated, deep-diving, fast-evading submarines with an extremely high PK (probability of kill) thanks to its inherent speed, its remarkably accurate seeker and its resistance to jamming and decoying. With the MU90 and the latest generation of active low frequency dunking and ship-mounted sonars, the French Navy has at its disposal the required tools to counter the diesel/electric submarines' enhanced operational flexibility. The MU90 remains in service on French Navy surface combatants, on the NH90 Caïman and on the Atlantique 2 maritime patrol aircraft which can carry up to six of them.

Armed with a Mk 46 training torpedo, this Flottille 34F Lynx is about to simulate a submarine attack. Such exercises were quite common for French Lynx aircrews.

    The Lynx could be equipped with a pintle-mounted 7.62mm ANF-1 machine gun to protect its mothership against asymmetric threats in contested waters. Alternatively, a sniper armed with a 0.5in (12.7mm) high-power precision rifle could be carried for drug interdiction, with the rifle securely fitted to a modular mounting for added stability. For self-defence, the helicopters were fitted with Saphir B0 decoy dispensers for flares to defeat infrared-guided surface-to-air missiles.

Limited Upgrade

To maintain the type's operational capability, ensure its military relevance in the short term and guarantee full compliance with the latest ICAO (International Civil Aviation Organization) rules and regulations, seven Lynxes went through a limited upgrade programme, the last update before the type was withdrawn from use. 'In 2017, the Lynx was, at last, fitted with an HF radio and with the associated antenna on the left-hand side of the fuselage,' explains Lieutenant Commander Christophe (surname withheld on demand), the head of 34F operations. 'They allowed us to talk to ships at very long distances. A new UHF radio with 8.33KHz spacing was adopted to support the Link 11 datalink functionalities and to communicate with civilian air traffic control agencies. Similarly, we were retrofitted with a Mode C/S/4 transponder. The sensor operator in the main cabin was equipped with a more modern workstation with a display that showed all data from the AIS, the L11 datalink and the Chlio FLIR turret.' The AIS (Automatic Identification System) is used to locate and identify ships at sea or at anchor, allowing maritime authorities to accurately track and monitor shipping. Data is automatically broadcast every 20 seconds (vessel identity code, position, speed, course, true heading) and every six minutes (radio callsign, name, type of ship and load, dimensions including draught, type of positioning system and its location on board the vessel, destination and estimated time of arrival).

The Super Frelon was well equipped for the anti-submarine role but was too heavy and too big for frigates. The type mainly specialised in SAR and in combat SAR and was eventually withdrawn from use in 2010.

The Sikorsky HSS was a successful aircraft in French Navy service. This aircraft is preserved at the entrance of BAN Lanvéoc-Poulmic.

While Royal Navy Lynxes were flown by a single pilot and an observer in the forward cockpit, French Lynxes were flown by two pilots with a crewman in the main cabin to operate the winch and sonar and take care of passengers and cargo.

By modern standards, the Lynx sonar suite was outdated.

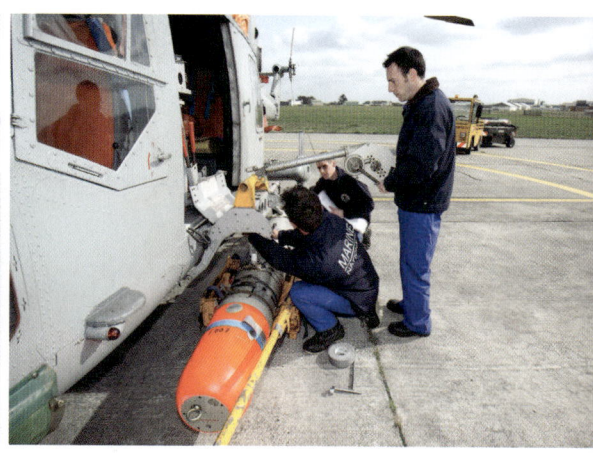

Above left: With a Royal Navy Lynx HAS8 hover taxiing in the background, an armourer pushes a Mk 46 torpedo towards a waiting Lynx.

Above right: A Mk 46 torpedo is being attached to a Lynx weapon pylon.

Left: The Mk 46 torpedo was the Lynx's weapon of choice from the helicopter's entry into service until 2010, when it was replaced by the much more modern MU90.

Above left: A Mk 46 training torpedo is being released from a Lynx. The parachute stabiliser has just begun to deploy to bring the torpedo into a correct angle for water entry.

Above right: After the firing, the training torpedo floats at the surface of the ocean, waiting to be recovered.

Left: A diver has jumped into the water to recover the torpedo. It will be brought back to BAN Lanvéoc-Poulmic where it will be serviced and made ready for re-use.

French Lynxes were often painted with special markings such as this cartoon which commemorates a deployment off Lebanon in 2008.

A French Navy Lynx hover taxies in at Lanvéoc-Poulmic in 2009. Noteworthy is the Royal Navy Lynx HAS3 in the background.

A Lynx overflies a French Navy semaphore near Brest. The French Navy operates a network of 58 semaphores that monitor all shipping off the French shores.

Above: In 2011, during the crisis in Libya, this Lynx lifts off from ASW frigate *Dupleix*.

Left: This photo, taken from the roof of the hangar, clearly shows how small the platforms are on F70 *Georges Leygues*-class frigates.

A Lynx harpoon locked into the deck steel grid to secure the helicopter.

Above: Noteworthy is the Chlio FLIR turret on the port side of the *Dupleix* Lynx. The Chlio is used to identify ships at stand-off distances.

Right: The *Dupleix* Lynx comes in to land back on board the frigate after yet another mission monitoring the whereabouts of any suspicious or hostile shipping during the 2011 crisis in Libya.

A Lynx being refuelled on the deck of *Dupleix* under the close surveillance of the flight deck director, an armourer equipped with a fire extinguisher and a fireman in heavy gear.

A fireman and a flight deck director having a chat between two sorties.

Lynx s/n 807 photographed on board *Dupleix*. It is about to be pushed inside the hangar at the end of an operational surveillance sortie off Libya in April 2011.

A Lynx undergoing checks after a sortie. French Lynxes were kept in pristine condition until the end, the maintainers devoting a lot of time and efforts to keep them serviceable.

This Lynx has its main rotor blades folded. The operation was purely manual. This was not a problem in good weather conditions, as in the photo, but much harder in bad weather.

A NH90 Caïman (left) and a Lynx share the ramp at Lanvéoc-Poulmic. The Caïman is a much bigger aircraft than its predecessor.

Lynxes were capable of operating in all conditions, day and night.

Up to two Lynxes could be accommodated on French anti-submarine F70-class frigates.

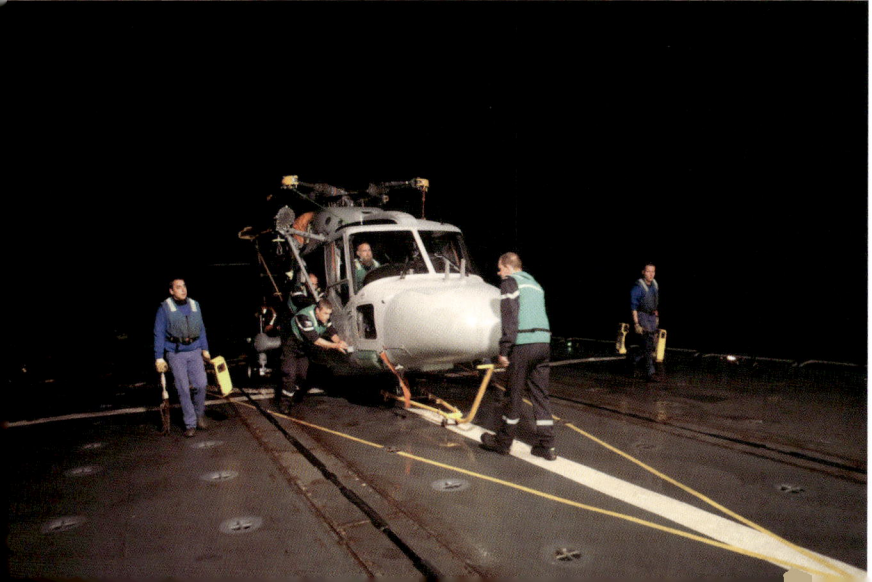

A Lynx being manhandled into the hangar at the end of a night training sortie on board ASW frigate *Latouche-Tréville*.

Above left: A DUAV-4 sonar transducer exposed on its maintenance cradle.

Above right: By modern standards, the sonar operator workstation looks totally outdated.

Below: The DUAV-4 sonar winch and motor exposed at Lanvéoc-Poulmic in September 2020.

Chapter 2
Lynx Testers

Right until the end, a dedicated French Navy unit, CEPA/10S, was responsible for all operational trials involving the Lynx.

The first Lynxes were delivered to Escadrille 20S of the CEPA (Centre d'Expérimentations Pratiques de l'Aéronautique Navale, the French Navy operational evaluation unit) at Fréjus-Saint Raphaël where the aircraft underwent an in-depth evaluation and where tactics were developed and refined. Until its disbandment in 1991, Escadrille 20S handled all rotary evaluation programmes. In the final years of the Lynx operations, the very same role was undertaken by Escadrille 10S of the CEPA, based in Hyères, also in the south-east of France. A team of CEPA/10S experts oversaw the trials and introduction of the MU90 torpedo in close cooperation with their colleagues at the Centre d'Essais en Vol, the French Flight Test Centre. Numerous separation trials and test firings were conducted from 2006 to clear the new weapon for service, including some trials in the Cazaux Lake, in the south-west of France.

The CEPA/10S also took an active part in the development of the Lynx's final upgrade, ensuring that performance levels matched the requirement. The unit kept a Lynx (s/n 267) operational at Hyères until nearly the end. Logistical and engineering support was provided by Lanvéoc-Poulmic, with the Lynx flight within the CEPA/10S being similar in many ways to a detachment at sea.

The CEPA markings are clearly visible on the door of this Lynx during the ceremony marking the 100th anniversary of the CEPA, at Hyères, in June 2016.

A beautiful line of rotorcraft at Hyères in 2016: a CEPA Lynx, Flottille 31F Caïman and US Marine Corps MV-22B Osprey. The nose of a Falcon 10MER can just be seen between the Caïman and the Osprey.

This modernised Lynx is fitted with a Chlio FLIR turret. Noteworthy is the temporary civilian registration number used for trial programmes.

The new MU90 torpedo adopted for the Lynx, Atlantique 2, NH90 Caïman and various classes of surface combatants, was tested by the CEPA.

Chapter 3
Flottille 34F and the Lynx

Flottille 34F was the last French naval squadron to operate the Lynx. The unit had flown the type for 41 years, from 1979 to 2020.

Since 2010 and the temporary disbandment of Flottille 31F (the unit reformed as a Caïman squadron in October 2012), Flottille 34F had been the only Lynx front-line operator in France. The unit's origins can be traced back to September 1974 when it was created at Fréjus-Saint Raphaël, in the south of France, to develop anti-submarine tactics using Alouette IIIs equipped with a magnetic anomaly detector. It initially deployed on board anti-submarine frigates *Duperré* and *Tourville*. In January 1975, the unit transferred to Lanvéoc-Poulmic, in Brittany, where it remained until the end. It received its first Lynx on 3 October 1979, and it began its transition from the Alouette III to the new twin-engine rotorcraft. Numerous detachments on board *Duperré*, *Tourville*, *Dugay-Trouin*, *De Grasse*, *Primauguet*, *La Motte Picquet*, *Cassard* and *Latouche-Tréville* were operated by the squadron with the Alouette III and then the Lynx.

Intense Activity

The final year of activity proved fairly intensive for Lynx aircrews and maintainers, with numerous detachments at sea on board *Georges Leygues*-class frigates *La Motte Piquet* (D645) and *Latouche-Tréville* (D646). Commander Chaput, Flottille 34F Commanding Officer, a very experienced pilot credited with nearly 7,000 flying hours (including 2,200 in the Lynx), says:

In August 2019, *La Motte Piquet* was engaged as the naval on-scene commander for the protection of the G7 summit in Biarritz, in the south-west of France. She acted as the command and control platform for all naval and air assets at sea, and over the sea. Two Lynx helicopters stood ready on board to help intercept any offenders infringing the temporary boating navigation restrictions that were being enforced for the duration of the summit. The frigate next participated in the Joint Warrior and Griffin Strike multinational exercises, off Scotland, before coming back to her Brest homeport in late October 2019. In late January 2020, she left Brest, in Brittany, for a deployment in the Arctic, a strategic area where tensions are rising. She operated alongside the *Charles de Gaulle* and came back to her homeport in incredible circumstances in April 2020, right in the middle of the Covid crisis.

Six *Rubis*-class submarines were delivered to the French Navy. They are the smallest nuclear attack submarines in the world. The six *Rubis*-class nuclear attack submarines will all be replaced by units of the new Suffren class. The *Suffren*, the lead ship, was recently delivered to the French Navy.

Although based in Toulon, in the south of France, nuclear attack submarines routinely deploy to the Atlantic to work with ballistic missile submarines and to train against ASW assets.

Flottille 34F and the two frigates also participated in traditional ASW missions to protect the ballistic nuclear submarines belonging to the Marine Nationale nuclear deterrence force, thus helping contribute to the overall credibility of the French nuclear deterrence policy. The Lynxes and the two warships also trained against nuclear attack submarines to hone their ASW tactics, while submarine crews perfected their skills in evading ASW forces. The activity sustained by Flottille 34F over the last year was intense, with each pilot logging 235 flying hours on average.

The art of Submarine Hunting

Flottille 34F aircrews devoted about 30 percent of their allocation of flight hours to ASW training. To hunt submarines, French Lynx helicopters were fitted with the DUAV-4, a rather old but still effective dunking sonar. Its range was not very long compared to the latest airborne sonars, but it nevertheless remained excellent at determining speeds and headings, and its ability to classify targets was outstanding thanks to good signal processing. This proved to be a crucial advantage to confirm the presence of a submarine prior to engaging it with a torpedo. Positive identification of underwater targets remains a challenging task due to the complexity of the marine environment; a whale produces a sonar return very close to that of a submarine under certain circumstances.

During CASEXs (Combined Anti-Submarine Exercises), Lynx aircrews trained in realistic conditions. While on board a frigate, the Lynx could be maintained at five-minute readiness, with the crew strapped on board the helicopter, or at fifteen-minute readiness, with the Lynx captain staying in the warship's operations room to keep an eye on the evolving tactical situation, and the co-pilot and the sonar operators resting in the ready room. When required, the Lynx could then take off to investigate a potential submarine contact and confirm her heading, speed, depth and, if feasible, identity.

When operating with the sonar dome in the water, the Lynx would typically hover at anything between 60 and 80 feet above the surface, depending on weight and sea state. To facilitate station keeping, the pilot could select a specific autopilot mode that would precisely maintain the hover above the cable and the sonar dome. Crews first lowered the sonar

Modernised Lynxes can be easily identified thanks to their white HF antenna fitted to the port side of the fuselage.

transducer to its maximum operating depth in order to obtain a bathy profile of the sea. This allowed them to accurately predict sound propagation speeds relative to depth and to know where the 'hostile' submarine was more likely to hide.

Against a real submarine during a CASEX, an expendable ESUS Mk 84 underwater sound signal system was manually dropped from the main cabin to inform the crew of the submarine that it had been fictitiously attacked by the Lynx. This type of system generates a coded acoustic signal easily intercepted by the submarine's listening devices.

Fully Operational Until the End

According to all interviewed pilots, the Lynx was, right until retirement, a very potent rotorcraft in terms of performance and power. Commander Chaput continues:

> The Lynx was fast, agile and reactive thanks to its rigid rotor head, with a cruising speed of anything between 120 and 150 knots depending on its configuration. Over recent years, French Lynxes had suffered from the fact that their tail rotor did not turn in the right direction. When Westland offered us to switch from older metallic blades to the newer BERP [British Experimental Rotor Programme] composite main rotor blades that provide increased performance and reduced maintenance requirements, the French Navy did purchase only one part of the kit for budgetary reasons: our Lynxes were therefore fitted with the BERPs, but not with the improved reverse direction tail rotor that turns in the opposite direction from the one we had. As a result, the aircraft sometimes proved difficult to control in yaw. The problem was exacerbated by the fact that the Lynx only had a small vertical tail: while the Panther and the Dauphin are fitted with a large vertical stabiliser effective enough to aerodynamically oppose the torque generated by the main rotor when flying in forward flight, this was not the case on the Lynx. Moreover, the Lynx was not equipped with modern aids for the pilots: while the Caïman proves extremely comfortable and easy to fly due to its advanced autopilot with countless modes, nearly everything had to be done manually on the Lynx as its autopilot was fairly basic.

Very Powerful and Agile Aircraft

Thanks to its two Gem 41-1 engines rated at 1150hp each, the Lynx was a very powerful helicopter. Lieutenant Commander Christophe (surname withheld on demand), the head of 34F operations, states:

> The level of power we had undoubtedly gave us confidence, especially in the hover, during sonar operations at night or in bad weather. Should an engine have failed, we could have easily performed a fly away on the remaining turbine, or even maintained the hover. I remember that a crew lost an engine during a rescue operation and still managed to complete the winching. We had a lot of power margin, and performance started decreasing only when temperatures soared above 35 degrees. With one Mk 46 torpedo and the sonar, we could operate at maximum take-off weight, with 780kg of fuel. With two Mk 46s, we could still take off with 580kg of kerosene. The more recent MU90 was heavier than the Mk 46 and we did carry only one, flying with 680kg of fuel which typically gave us 1h 40min of endurance.

The Lynx was also highly praised for its agility. Lieutenant Commander Christophe continues:

> The rigid rotor head allowed us to fly the aircraft extremely accurately, and the Lynx was incredibly responsive. It was significantly more agile than the Dauphin and even than the Gazelle. It was so manoeuvrable that we had to be careful not to exceed the g-limit set at 1.4 g, not that easy in an aircraft with no g-meter. We could still operate with 30 knots of winds on the sides or from the back, and with up to 60 knots head on. This was a crucial advantage when fast roping commandos onto a ship or when winching survivors out during rescue operations.

Retro Markings

To commemorate the withdrawal of the Lynx and honour former aircrews and maintainers, Flottille 34F decided to paint two airframes in retro markings. 'The project was approved by the Naval Aviation Command and by the Aeronautical Maintenance Directorate,' says Commander Chaput. He continues:

> We chose the two aircraft that had the largest amount of remaining flight hours on their airframes. For Lynx No 622, we adopted the very same white/grey-blue paint scheme with a black nosecone that was

The black nose paint that was adopted for the two Lynxes with retro markings was not compatible with radar transmissions.

used at the beginning of the Lynx operational career. Lynx No 273 was adorned with the overall grey-blue scheme with a black radome adopted a couple of years later. Maintenance experts had to look for the exact same paints used at the time. They did not have any problem sourcing the white and the grey-blue paints but were unable to source the black paint that had been approved for the Lynx nosecone. The problem was that the new black paint is not approved for nosecones as it blocks radar waves. As a consequence, the black nose radome had to be replaced with a standard grey one for operational missions. Luckily, we had a few spare grey nosecones which gave us the required flexibility to swap noses around as necessary. Two weeks were needed to paint each airframe. The work was performed by third-line painting specialists with additional support from Flottille 34F personnel. I am very happy that the project attracted a lot of support from higher echelons and aroused keen interest within the Lynx community.

Lynx inventory

In early 2020, Flottille 34F still had eight aircraft, but the number would quickly diminish. By mid-June, only s/n 272, 273, 621, 622 and 806 remained. S/n 267 was still at Hyères with the CEPA/10S the French Navy operational evaluation unit. 'Lynx availability varied significantly from one month to another all along its last year of operations,' admits Commander Chaput. 'We found a number of cracks on various aircraft which led to the decision to retire early one of our Lynxes because the cracks were so severe they were found to be beyond economical repair for just a few months of service left. Thankfully, we had some margin as the other choppers had enough remaining hours on their airframes to accept the extra load in terms of flight hours needed to carry out the operational missions requested by the Marine Nationale.' By the end of July, only four Lynxes remained in flyable condition: s/n 267 (ex-CEPA, by then transferred to Flottille 34F), 272, 273 and 622. With these four airframes, the unit officially stayed operational until 3 August 2020. For the final demo conducted on 4 September 2020, only s/n 267, 273 and 622 were still in flyable condition, and only 273 and 622 flew.

The French Navy maintained extremely good relations with the manufacturer of the aircraft, at Yeovil, throughout its remarkably long life.

Lynx aircrews routinely exercised against French and foreign submarines. The crews are all trained to an extremely high standard.

A trio of Flottille 34F Lynxes share the ramp at Lanvéoc-Poulmic in June 2019.

The two specially painted Flottille 34F 'retro' Lynxes in a playful mood off Brittany in June 2020. The retro markings adopted by Flottille 34F in the final months of operation have proved highly popular within the Lynx community, attracting a flow of positive comments.

Lynx playing hide-and-seek behind the Rocher du Lion (lion's rock) off the tip of the Crozon Peninsula, in Brittany.

Above: Lynx s/n 273 in the hover at very low level. The green patches on the side of the airframe and on the undercarriage sponsons hide the emergency flotation gear that would be triggered in case of a ditching.

Right: This Lynx manoeuvres hard, initiating a tight turn to the left. The Lynx is renowned for its agility. Noteworthy is the harpoon and the hatch for the sonar.

A classic rotorcraft that will be sorely missed. The Lynx was well liked by French aircrews and engineers alike.

A three-ship formation manoeuvres at low level close to the Pierres Noires (black stones) lighthouse, iconic flying machines overflying an iconic landmark.

The Lynx is well known for its exceptional agility and power margin.

This formation shows the three main camouflage schemes used by the French Navy during the Lynx's career, from the oldest one on the leading aircraft to the most recent on the helicopter the furthest away.

This photo shows how the French Navy toned down its aircraft markings step-by-step, making them progressively better camouflaged and less conspicuous. The latest two-grey camouflage is incredibly effective over the sea.

This Lynx shows its DUAV-4 dunking sonar to good advantage. The sonar dome could be lowered hundreds of feet down into the water in just a couple of minutes.

Lynx s/n 622 comes into a hover next to the *Jaguar*, a Brest-based French Navy training ship.

Lynx 622 seen head on, with its radar-compatible grey nose. Readers should note the position of the main wheels optimised for deck landings.

Right: The grey nose seems a little bit awkward on Lynx s/n 622. The aircraft is fitted with Saphir B0 decoy dispensers on the upper surfaces of the undercarriage sponsons.

Below: Lynx s/n 622 cruising at 10,000 feet. Thanks to its enormous power, the Lynx is at ease in all sorts of environments.

Lynx s/n 622 fires a salvo of flares from its Saphir B0 launchers. Lynx self-defence capabilities were progressively expanded to address a whole range of new threats, including shoulder-launched, short-range, infrared-guided surface-to-air missiles.

An MU90 training torpedo is removed from its transport container by two Flottille 34F armourers.

Fitting an MU90 torpedo to a Lynx weapon pylon was incredibly quick. This helped reduce turnaround times and boost the aircraft's operational efficiency.

The MU90 is one of the most lethal and most recent airborne torpedoes in service anywhere. Training MU90 torpedoes are painted orange.

Chapter 4
French ASW Frigates

French Navy Lynx helicopters mainly operated from the three F67 *Tourville* and seven F70 *Georges Leygues*-class anti-submarine frigates, warships that proved highly successful in Marine Nationale service.

The F67 warships (*Tourville*, *Duguay-Trouin* and *De Grasse*) that entered service between 1975 and 1977 initially deployed with Alouette III light helicopters before switching to the Lynx in the late 70s/early 80s. The F70 frigates (*Georges Leygues*, *Dupleix*, *Montcalm*, *Jean de Vienne*, *Primauguet*, *La Motte Picquet* and *Latouche-Tréville*) were designed right from the beginning to operate with the Lynx. These ten ships formed the backbone of the French ASW force during the last decade of the Cold War, until the mid-2010s. Today, only *Latouche-Tréville*, the last F70 unit, remains in service. She will decommission in 2022, after 32 years of service.

Bird's eye view of F70-class frigate *Latouche-Tréville* as she sails off Brest in July 2020. By modern standards, the frigate does not look too stealthy.

Active Variable-Depth Sonar

With the F67 and F70 frigates, the French Navy pioneered the use of towed, active, variable-depth sonars (VDS) to detect and track hostile submarines at very long distances in conjunction with rotorcraft and other assets such as maritime patrol aircraft and French and allied nuclear attack submarines. The VDS concept had already been tested on seven earlier type 53/56 destroyers (*La Galissionnière*, *D'Estrées*, *Maillé-Brézé*, *Vauquelin*, *Casabianca*, *Guépratte* and *Duperré*), but only *LaGalissionnière* could carry an Alouette III. *Duperré* was refitted to a more advanced standard, with a larger platform, a bigger hangar and with a dedicated helicopter securing and handling system. She did eventually carry a single Lynx, paving the way for the introduction of the F67s and F70s. Air-defence destroyers *Suffren* and *Duquesne* also had a VDS, but no helicopter. Introduced in 1973, frigate *Aconit* was optimised for ASW ops but with no helicopter; a choice difficult to understand at a time when rotorcraft had already proved their worth. Only one ship was produced before the project was abandoned, with the Marine Nationale switching to the F67 instead.

Increased Operational Capabilities

When F67 and F70 classes appeared, they brought significantly increased operational capabilities, especially when they carried a detachment of two Lynxes. Each frigate and her helicopters could act as a deadly hunter-killer team, either as part of a larger task force, or alone when conducting operations autonomously. Their ASW gear was constantly upgraded as large investments were made to progressively refine the sonar technology to match the emerging threat as quieter and deeper diving submarines entered service all over the world. As a result, Thales is today considered a world leader in active/active low frequency VDS.

Tactics were also progressively improved as frigate officers and helicopter crews learned how to exploit to the full their respective systems and how to master ASW tactics in both blue waters and littoral environments, often cooperating with Atlantique 2 maritime patrol aircraft.

Similar, but Different

Although F67 and F70 frigates look rather similar, the F67s are in fact much bigger, with a displacement of 6,100 tonnes full load, compared to 4,900 tonnes for the newer, more compact F70 vessels. They can be easily distinguished from the F70s thanks to their two 100mm gun turrets at the bow, the shorter F70s being equipped with only one gun turret. The *Tourville*-class vessels were powered by boilers and steam turbines and offered excellent sea-keeping qualities. Like all warships of their generation, they required a large ship company, however, with a crew of about 300 officers, NCOs and sailors. This was not a real problem when conscription was still in place but became a clear financial disadvantage (some even say burden) when France switched to all professional armed forces from 1996. This is one of the reasons that led to the premature retirement of *Duguay-Trouin* in 1999. *Tourville* and *De Grasse* had both been modernised and soldiered on until 2011 and 2013 respectively.

The later F70s were more automated than their predecessors, with a more modern CODOG (combined diesel or gas) propulsion system divided into two SEMT Pielstick diesel engines and two powerful Rolls-Royce Olympus turbines. Thanks to the automation of their propulsion system and of their mission system/combat information centre, their crew was reduced to about 240 personnel.

Like their predecessors, the F70s were very stable in all weathers and extremely seaworthy, a major advantage when towing a massive VDS in heavy swell or when launching and recovering helicopters. As a result, they were highly effective tools in the ASW role, as they could remain operational even when the weather conditions worsened. They are also remarkably robust vessels that have proved very reliable over the years.

F67-class frigates were bigger than their F70 derivatives. They could be easily identified with their two 100mm gun turrets forward of the bridge. F70-class frigates have only one gun.

Multirole Frigate

The temporary disappearance of the Russian submarine threat at the end of the Cold War and the rise of tensions in Bosnia and in the Middle East led to a fast-growing number of deployments to the Adriatic, the Persian Gulf, the Oman Sea and the Arabian Sea. The ASW frigates were then increasingly used as multirole platforms, conducting autonomously a wide range of missions, including intelligence gathering, embargo enforcement, drug interdiction, military presence at sea, escort and protection of civilian and military high-value assets such as tankers transiting through the Strait of Ormuz and sea control operations as part of the global war on terrorism. F67 and F70 frigates proved well adapted to all those missions. They often escorted carriers *Clemenceau* and *Foch* involved in operations in Bosnia and Kosovo. From 2000, they routinely deployed with the *Charles de Gaulle* carrier strike group to provide deterrence and protection against hostile submarines. They also remained focused on ASW and kept on training with, and against, French and NATO submarines. To reduce costs, the number of Lynxes in service had been slashed, and the frigates usually deployed with only one Lynx instead of the two carried at the peak of the Cold War; though some ships did carry two helicopters for given missions.

Tourville was the first of the three F67-class frigates. She entered service in 1975 and was withdrawn from use in 2011, after 36 years of sterling service.

Tourville seen during an underway replenishment at sea. She is sailing immediately astern of the fleet tanker.

F70-class frigate *Dupleix* photographed off Libya during the war in 2011. She paid off in July 2015, after 37 years of service.

'Tail Wagging' Frigates

F70-class frigates are well known by French Navy rotary pilots for their ship motion in yaw. Commander Chaput, the last Flottille 34F Commander, said that:

> The *Latouche-Tréville*, like all her *Georges Leygues*-class sister-ships, is fitted with stabilising ailerons on each side, port and starboard, below the waterline. Thanks to gyroscopic information, they are hydraulically activated to stabilise the ship in roll in a strong sea swell. As soon as the system detects a movement about the roll axis, the aileron on the side onto which the frigate is leaning is activated to counter that rolling motion. As it moves up or down, the aileron instantly increases the hydrodynamic drag on that side. As a result, the ship constantly yaws from one direction to another. It is a phenomenon in many ways similar to the reverse yaw/induced roll that affect fixed-wing aircraft. As we say in the French Navy, F70-class ships are 'tail wagging frigates'. It sometimes proves very annoying, especially at night in rough seas, even with night vision goggles, because our visual references keep on moving all the time. The more recent *La Fayette*-class stealth frigates are equipped with stabilising ailerons coupled with the rudders. This is much more effective because the rudders are programmed to counter the yaw created by the ailerons. Even though they are smaller, the stealth frigates prove more stable than the earlier F70s for the helicopter pilot on short finals.

With only *Latouche-Tréville* remaining, the replacement of the ten F67 and F70 frigates is now almost over. Six of the eight new 6,000-tonne *Aquitaine*-class multi-mission stealth frigates are already in service, with the last two to be delivered in 2021 and 2022. They will be followed by five 4,460-tonne *Amiral Ronarc'h*-class defence and intervention frigates from 2023 to 2029. These 13 new generation vessels will all feature robust ASW capabilities.

State-of-the-art air-defence destroyer *Forbin* followed by *Tourville* in the Indian Ocean in January 2011. This photo clearly shows how dated the *Tourville*-class ships already looked at the time compared to *Forbin*.

Left: Nuclear carrier *Charles de Gaulle* is always escorted by at least one air-defence destroyer such as *Forbin* and one anti-submarine frigate (here *Tourville* in early 2011).

Below: F70-class frigate *Jean de Vienne* sailing in close formation with Type 23 frigate HMS *Northumberland* as both vessels escort carrier *Charles de Gaulle* during Exercise *Corsican Lion* in October 2012.

Dupleix's 100mm gun pointed at an imaginary target for the photo. The 100mm gun can engage both surface and airborne targets with deadly accuracy. It can deliver 78 rounds per minute.

Some of the F70 frigates were equipped with 30mm gun turrets for self-defence.

Impressive view of F70-class frigate *Primauguet* being towed under the Recouvrance Bridge, in central Brest, after coming out of re-fit in dry dock. The clearance between the tip of the mast and the bridge is just a few metres, and the tide has to be carefully taken into account.

The *Latouche-Tréville*'s combat information centre is a mix of old and new technologies. The frigate sensors and software have been regularly upgraded, but some of the displays seem outdated.

The *Latouche-Tréville*'s engine control room during the COVID crisis, with sailors wearing face masks. The diesel engines and the turbines are remotely operated.

Below left: One of the two enormous diesel engines that propel the *Latouche-Tréville* frigate.

Below right: *Latouche-Tréville*'s electric energy is provided by massive diesel alternators, all widely separated to ensure much needed battle damage redundancy.

Above: At the rear of the F70-class frigates is installed a powerful towed active sonar used to track submarines at very long distances.

Right: The impressive machinery of the DUBV-43 variable-depth sonar is clearly visible on this photo of the *Latouche-Tréville*.

The DUBV-43 variable-depth sonar about to be immerged. The reeling cable is 700-metre long and the sonar weighs over 10 tonnes!

Latouche-Tréville sailing in spectacular scenery off Brittany in July 2020.

Another impressive view of *Latouche-Tréville* sailing in close proximity to rocks off Brittany. In such conditions, the duty crew on the bridge and the officer of the watch have to be extremely careful to avoid a disaster… and a major embarrassment for the Navy! Thankfully, they are trained to very high standards.

Left: F70-class frigates are equipped with a whole range of sensors to detect and track airborne, surface and underwater threats. Here, *Latouche-Tréville* heads towards the south alongside the coastline after leaving Brest.

Below: F70-class frigates are inherently agile. The two massive radomes above the forecastle hide Syracuse satellite communications antennas that allow the ship to remain in contact with higher echelons (and vice versa).

Above left: For self-defence, the *Latouche-Tréville* is equipped with the Crotale surface-to-air missile launcher that can engage aircraft and anti-ship missiles.

Above right: Four MM40 Exocet missile canisters loaded in their launchers. A maximum of eight Exocets can be carried in wartime, although four is the normal peacetime load.

Right: *Latouche-Tréville*, the last of the French Cold War frigates, will be withdrawn from use in 2022, after 34 years of service. At the time of writing, *Latouche-Tréville* was due to enter re-fit in early 2021 for the last time before retirement in mid-2022.

Four live MU90 combat torpedoes and one training torpedo stored in the *Latouche-Tréville* rear torpedo weapon hold, close to the helicopter hangar.

Left: An internal crane is used to lift the MU90 torpedo from its storage cradle to its transport cradle.

Below: *Latouche-Tréville* at sea off Brest. The F70-class frigates have proved to be dependable and reliable vessels. They offer excellent sea-keeping qualities, a major advantage when operating with a towed active sonar.

Chapter 5
Special Forces Lynx

Lynxes and special forces commandos have long participated in common operations, helping create a very effective force for a whole range of missions.

A dedicated command of the French Navy specialises in special forces operations. It oversees the missions of commandos trained to very high standards. The history of French commando units can be traced back to World War Two when Free French units, trained by the British Armed Forces, operated alongside their Commonwealth counterparts. Today, the French Navy special operations command is split into seven commando units, six of which are stationed in Lorient, Brittany:

- Commando Hubert, a combat swimmer/combat diver outfit based in Saint-Mandrier, in the south of France;
- Commandos Jaubert, de Montfort, de Penfentenyo and Trépel, all four of them focusing on traditional special forces missions, including reconnaissance, assault, maritime counter-terrorism and sniping missions;
- Commando Kieffer, which specialises in UAVs, CBRN, electronic warfare and assault dogs;
- Commando Ponchardier, which provides command and control assets, signal cells, fast-attack craft and various other skills.

Lynxes in Support of Special Forces

The Lynx community has long participated in the support of French Navy special forces. Lanvéoc-Poulmic, where the Lynxes of Flottille 34F were based, is less than an hour's flight time away from Lorient, where most of the commandos are stationed. This proximity led to close ties being established between the squadron and the various commando units. Flottille 34F was not listed as a special forces unit, as it was clearly focusing on ASW and on the protection of nuclear ballistic submarines. Most

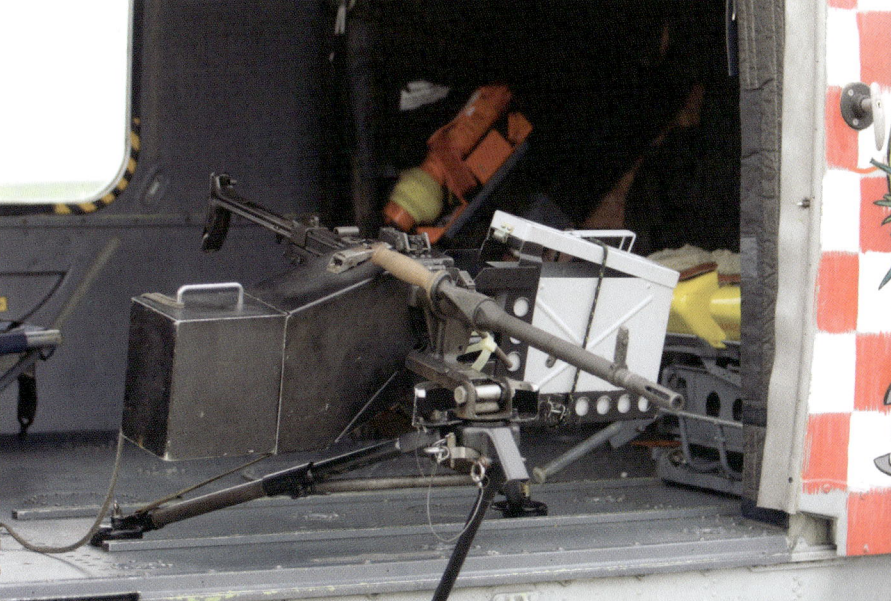

A pintle-mounted ANF-1 7.62mm machine gun photographed on a Flottille 34F Lynx.

Above: The nimble, powerful and agile Lynx excelled in the special forces support role, either as a transport/assault platform or as a fire-support platform with snipers or a machine gun on board.

Left: The main drawback of the Lynx when operating with special forces was the small size of the cabin. Here, commandos are being airlifted prior to a simulated airborne assault.

All French commandos, be they from the Army, the Air and Space Force or the Navy, are trained in fast roping techniques. They have to maintain currency, training regularly with helicopters.

of its aircrews were trained in supporting special forces, however, and knew the specific procedures developed and used by the commandos. They routinely trained together to maintain their respective operational qualifications. Occasionally, Flottille 34F aircrews would also train with Army or Air Force special forces operators as part of increasingly joint operations.

The Lynx served as a fire-support and an assault platform. It could also perform medical evacuations or resupply flights in support of special operations. In the special forces role, its agility and power were obvious advantages, as the pilot could instantly react to the sudden appearance of a pop-up threat. Its main cabin was cramped, however, limiting the load to only five or six fully equipped commandos who could be inserted in various ways, using fast roping or abseiling techniques, or even parachuting from higher altitudes during an insertion for example. The Lynx was considered a stable firing platform for snipers armed with McMillan, Barret or PGM Hécate 12.7mm precision rifles, or for fire-support missions with a pintle-mounted ANF1 7.62mm machine gun.

Night Vision Goggles

French special forces mainly operate under the cover of darkness, and they need their supporting rotary assets to be trained at night too. The French Lynx community started flying with night vision goggles (NVGs) decades ago, fast gaining valuable experience and developing new tactics and techniques to exploit to the full the new capabilities offered by the goggles. Until the end, a large number of flying hours were dedicated to night-time missions with NVGs.

Maritime counter-terrorism figures among the other key missions of the Marine Nationale, with its helicopters playing a crucial role carrying assault teams and providing fire support to seize a vessel by force and liberate hostages. The Lynx remained available for such a mission right until the very end. It should be noted that skills developed for special ops were routinely used within the Lynx community for 'Narcops' (narcotics operations/drug interdiction) at sea, when stopping a go-fast and/or seizing a large vessel was required.

Most special forces ops are expected to be conducted at night and Flottille 34F aircrews are perfectly at ease flying with night vision goggles. Here, a team of commandos walk towards a waiting Lynx at Lanvéoc-Poulmic prior to a night training sortie.

Fast roping is considered a basic skill for all special forces in France. Noteworthy is the short-range radio fitted to the body armour. A longer-range radio-set is also carried in the backpack.

Left: Gloves don't last very long during this kind of training. The rate of descent is adapted by applying more or less pressure on the rope with hands and feet. More pressure is required to slow down when approaching the ground.

Below: During fast roping, hover height is adapted to reduce the amount of time required for the commandos to carry out their descent. Here, there were no obstructions so the Lynx remained low.

Above: A French Navy Commando displaying typical combat gear. He is armed with a HK416 5.56mm assault rifle equipped with a laser marker.

Right: Two commandos about to be recovered by a Lynx for the next training scenario. One of them has painted his HK416 rifle to enhance his camouflage. They belong to Commando Kieffer, a unit created in 2008 for highly specialised roles such as command and control, UAV operations, CBRN warfare, de-mining and electronic warfare.

The Commandement des Opérations Spéciales, the French special operations command, expends a lot of effort training Army, Air and Space Force and Navy commandos to very high standards.

Above: At the end of the training exercise, engineers, aircrews and commandos pose for the photo on the Flottille 34F flight line, at Lanvéoc-Poulmic.

Left: The Lynx will be missed by special forces, especially as a powerful sniper and machine-gun fire-support platform for maritime counter-terrorism operations.

This photo clearly shows how small the Lynx cabin really is. No more than four or five fully equipped commandos and a dispatcher can be realistically carried for operational missions.

Chapter 6
Lynx SAR Missions

During their distinguished career, French Lynxes performed countless search and rescue (SAR) missions, saving hundreds of lives, often in difficult conditions.

French Navy helicopters share coastal SAR duties with other rotary assets belonging to the French Air and Space Force and with the Sécurité Civile. They are spread in order to cover the entire French coastline, in the North Sea, the Channel, the Atlantic Ocean and in the Mediterranean Sea, as well as in various French dependencies in the West Indies, French Guyana, the Indian Ocean and the Pacific.

Five main helicopter types have been used for SAR by the Marine Nationale over the last 20 years, split into 'heavies' (Super Frelon, EC225 and NH90 Caïman) and light helicopters (Dauphin and Lynx). The SA321G Super Frelon was the main SAR asset in Brittany for nearly four decades, this three-engine helicopter earning an enviable reputation of efficiency and reliability. Very stable and big enough to evacuate the entire crew of a merchant vessel in one go, the huge rotorcraft proved ideal for rescues in adverse weather conditions. The Super Frelon was seconded from Lanvéoc-Poulmic by a single Dauphin and/or by a Lynx that could be drafted in as primary or secondary responder, depending on the situation. For example, it was sometimes more cost effective to scramble the Dauphin or a Lynx to winch a single fisherman out of a small trawler. The classic Super Frelons were withdrawn from use in 2010 and temporarily replaced by two EC225s that had been purchased as gap fillers. These state-of-the-art aircraft allowed French Navy aircrews to gain experience flying a modern helicopter fitted with the latest generation of systems, including multifunction displays. The two aircraft transferred to the Armée de l'Air to perform other roles as the number of NH90 Caïmans increased. The Caïman has since become the primary SAR asset but is due to be replaced in that role from 2022 by H160 helicopters leased to fill the gap until the arrival of the first H160M Guépard in 2028. The NH90 will then focus on combat missions.

A Lynx photographed from the bridge of training ship *Jules* during a SAR winching exercise in June 2019.

A Major Oil Slick

Within the Lynx community, rescue missions were always popular as they proved extremely challenging and rewarding. Commander François Chaput, Flottille 34F Commanding Officer, has agreed to share some of his experience, describing in great detail two gratifying rescues, the first one from Lanvéoc-Poulmic, and the second one from a frigate in the Mediterranean Sea. In December 1999, after the structural failure of her hull, oil tanker *Erika* broke in two off Brittany in gale force winds, triggering a massive rescue effort.

A Super Frelon had launched as the first responder but its winch had suffered a malfunction, and I was scrambled in a Lynx to bring to safety the last crew members still on board the oiler. Weather conditions were appalling, to say the least, with a 40-knot wind and a huge swell. Thankfully, we did intervene in broad daylight. By then, the ship's hull had split into two, with the stern and the bow widely separated. The survivors were all on the stern that seemed very stable, as if planted in the water. We started winching out *Erika*'s crew members. I had decided to keep my rescue diver inside the Lynx as the Super Frelon's diver had remained on board the stern to organise and supervise the rescue operation. After a while, my winchman told me it was time to go, and I answered that we still had enough power margin to continue winching crew members out. He was very surprised and asked me to have a look in the cabin. I could not believe what I saw: our Lynx's rather small cabin was completely full, with six survivors on top of the winchman and diver. I better understood his request, and I immediately performed a fly away and headed towards Saint-Guénolé harbour, the nearest drop zone where we unloaded the survivors directly onto the harbour. Then, I flew back to what was left of the *Erika* to winch out another two survivors and the Super Frelon's diver. The stern eventually sank a couple of hours later. In such demanding

The Super Frelon was an extremely stable platform in strong winds. Countless daring rescues were undertaken by Super Frelon crews, often in severe weather.

conditions, we do really appreciate flying such a powerful, agile and responsive helicopter as the Lynx.

The *Erika* disaster triggered a major oil slick that would require a massive cleaning operation on literally hundreds of kilometres of French coastline in what is considered one of the worst environmental catastrophes in Europe in recent years.

Rescuing a Belgian Skipper off Crete

A year and a half later, Commander Chaput was involved in another daring rescue off Crete, in the Mediterranean Sea. He recalls:

> I was on board frigate *Tourville*, north of Crete, on our way to a scheduled port visit to Suda Bay, in Crete. At about midnight, we received a call from the Maritime Rescue and Coordination Centre, in Piraeus, informing us that a small catamaran had declared an emergency. She was caught in one of these violent storms that sometimes occur in the Med. Her sails had been blown away and her engine had broken down. We immediately started planning, refining our fuel and weights calculations. Needless to say, we were really disappointed when the MRCC Piraeus called back to tell us that a merchant vessel sailing nearby had been diverted to help. We were like caged lions, ready to launch for a very promising mission when everything was called off. We went back to our respective cabins but had difficulty finding sleep as the adrenaline was still pumping. A couple of hours later, in a dramatic reversal of situation, we are called by the *Tourville*'s captain: the MRCC had asked us to intervene because the rescue attempt by the merchant vessel had gone wrong, with only three of the catamaran's passengers and crew members plucked out of the water via a pilot ladder. The sudden fall of the catamaran's mast had thrown into the water two hapless crew members.

With lives at stake, the crew hurried. The Lynx was launched in a very short amount of time.

> We transited south, through the mountains in Crete. Not that easy in a totally unknown area, without any night vision goggles. As soon as we were in radio contact with the merchant vessel, we were told they could hear the survivors calling for help but that they could not see them. Without wasting any time, I ordered our rescue diver to get ready. In the pale light of dawn, the first survivor was quickly located but he was apparently unconscious. In record time, our diver jumped into the water and immediately positioned the harness strap under the arms of the survivor. The loadmaster in the back soon began winching them both up. Unfortunately, the unconscious victim started sliding down through the harness because of his wet clothes, with the diver and the loadmaster unable to stabilise him.

Disaster struck when the poor sailor fell back into the sea.

> Our diver reacted instantly and, in literally no time, he miraculously found the victim again. They were both winched out of the sea again. Same cause, same result: he slowly started sliding downwards again through the harness… He was a really big guy, you know, very heavy, with wet clothes that retained water that added to his own body weight. The diver and the loadmaster were

doing their best to stabilise him, but they were struggling, totally stuck and unable to pull him into the cabin. As I was guessing he would soon fall back again, I hovered as low as I dared above the wave tops in an effort to cushion his fall. My co-pilot saved the situation. He suggested going to the back to help the diver and the loadmaster drag the survivor into the cabin. The problem was he would have to unstrap, and we did not have a spare harness. Rather risky with the right cabin door wide open... As the aircraft commander, I gave it a quick thought and eventually agreed. Did we have a choice? I do not think so. He unstrapped and crawled from his seat in the cockpit to the cabin, taking great care not to accidentally cut off the fuel controls that would instantly have starved the turbines. With his help, the diver and the loadmaster managed to drag the victim to safety.

The second survivor was rapidly spotted thanks to his red Mae West. 'He was conscious and was quickly winched up with no drama,' continues Commander Chaput. 'By then the diver was already trying to resuscitate the big guy. He managed to empty his lungs, and the victim, a Belgian skipper, was brought back to life after a short heart massage with mouth-to-mouth resuscitation. I am still in contact with him to this day.'

French Navy helicopter divers are experts in a wide range of skills, including first aid and medical evacuation.

Above left: Two EC225s temporarily replaced the Super Frelon, with the Lynx or the Caïman remaining available as back-ups.

Above right: Even though the Super Frelon was a very big helicopter, it proved incredibly agile, as confirmed by this photo taken in April 2010 off Brittany.

Right: Although the Lynx never was a primary SAR asset, it often operated alongside, or as reinforcement, to dedicated rotorcraft such as this Dauphin photographed in the background.

Below: Until 2010, the Super Frelon was the primary heavy SAR helicopter in France. This Flottille 32F aircraft was photographed a couple of weeks before its retirement.

The *Jules* was specially ordered in 2015 by the French Navy for helicopter winch training. She entered service in 2017 and is based at Lanvéoc-Poulmic, in Brittany. She has since proved so successful that a second, albeit smaller, boat has been ordered for winch training at Hyères, in the south-east of France.

Left: Most of the time, French Lynxes were flown with a rescue winch, meaning they were quickly available for a SAR mission should the need have arisen. The Lynx was equipped with a winch capable of lifting 600lbs (272kg), or two people at a time.

Below left: Lynx s/n 265 maintains a perfect relative hover above the stern of training ship *Jules* in June 2019.

Below right: One of the two EC225s was detached to Cherbourg, in Normandy, to provide SAR coverage in the Channel. The two EC225s nearly always flew with external fuel tanks bolted on both sides of the fuselage.

Above left: Two rescue divers prepare a litter for a winching/medical evacuation exercise with a Lynx helicopter.

Above right: A helicopter rescue diver lifts a litter off the deck of the *Jules* training ship during a SAR exercise.

Right: Physical strength is one of the qualities required to be a rescue diver. Here, a diver helps guide a litter onto a Lynx.

Below: An instructor winch operator photographed in the rear cabin of a Lynx. She is equipped with the new British-sourced Alpha helmet that now equips all helicopter aircrews in France, be they from the Army Aviation, the Air and Space Force or the Naval Aviation.

The Société Nationale de Sauvetage en Mer, the French equivalent to the British RNLI, also routinely trains with French Navy helicopters. Here, one of their lifeboats participates in a medical evacuation exercise with a Lanvéoc-Poulmic-based chopper.

Left: A rescue diver demonstrates the typical equipment used for SAR missions. Three air bottles are carried to give the diver the possibility to perform a subaquatic investigation; for example, inside an overturned sailboat to check that nobody is stranded inside the hull.

Below: French Navy rescue divers typically log about 80 to 100 flying hours per year.

Above: The Lynx winchman will soon pull the litter into the helicopter's main cabin. Easier said than done... This time the litter is empty. This is a totally different story with a victim tied to the litter.

Right: A litter ready to be airlifted from a lifeboat during a routine training drill in 2012.

Below left: NH90 Caïmans have become the main heavy SAR asset since the EC225s were transferred to the French Air Force. They will soon be replaced with leased H160s.

Below right: A NH90 Caïman being refuelled at Lanvéoc-Poulmic in 2017. This aircraft is the duty SAR helicopter. Should the need have arisen, a Lynx could have been called to help.

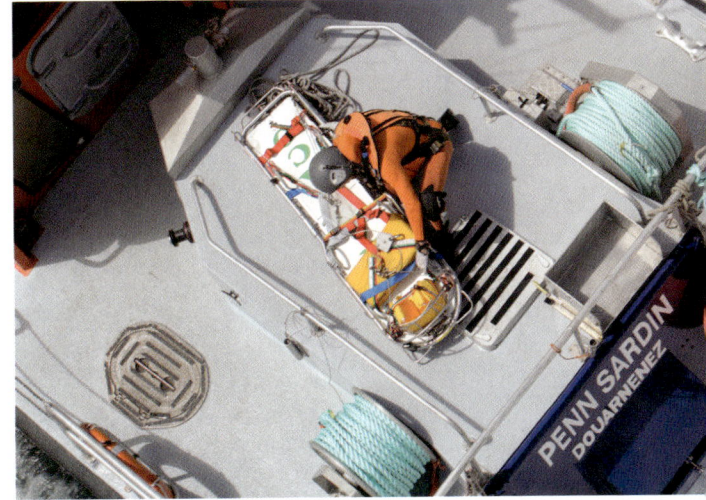

Chapter 7

Maintaining a Proven Rotorcraft

The Lynx could not have flown without the support of highly motivated and highly qualified teams of Marine Nationale engineers. They spared no effort keeping the Lynx in pristine condition until the end.

In the last year of Lynx operations, no fewer than 56 maintainers were dedicated to the maintenance of Flottille 34F's remaining rotorcraft. Lynx maintenance was split into 25, 50, 150, 200 and 400-hour inspections, with additional weekly and yearly checks, all carried out at Lanvéoc-Poulmic, at squadron or naval air station levels. Every three and nine years, depot-level inspections had been undertaken at Cuers-Pierrefeu, in the south of France, where most Marine Nationale fixed-wing aircraft and helicopters undergo third-line maintenance. The last nine-year check, on Lynx s/n 267, was performed there in 2019.

Remarkably Strong Airframe

The Lynx has proved to be a very strong aircraft, but the type was ageing, and some recurring technical problems had surfaced over the course of the last few years. The Lynx easily handled operations at sea and its airframe had been optimised to resist corrosion. The French Navy's main problem was that Leonardo (previously known as Westland, and then AgustaWestland) had discontinued the production of some of the Lynx components, and finding spares had become a real challenge. For instance, sourcing undercarriages had become a problem. Avionics were old, and aircraft suffered from a high number of electrical connector issues. To increase availability, helicopters standing ready for SAR missions were

At the end of its career, as the rotorcraft was ageing, French Navy engineers spent an increasing number of hours maintaining the Lynx for each flight hour logged.

parked in a humidity-controlled hangar, with positive results. The fleet did not have much difficulty soldiering on until retirement date because, in the course of the last depot-level inspections, modifications had been made to ensure that the airframes could clock 8,000 flying hours instead of the initially approved 7,000.

For the French Navy, the Lynx was not unlike the proven Alouette III: engineers had decades of experience on type and could solve all the airframe problems that cyclically appeared. The Lynx was a very powerful helicopter, with the main rotor inflicting a lot of torque on the fuselage while the tail rotor also created considerable lateral forces on the tail boom. As a result, French maintainers regularly discovered cracks on the frame, at the junction between the forward fuselage and the tail boom. Working within Flottille 34F proved very instructive and young maintainers straight from engineering school were still being posted to Flottille 34F until nearly the end so that they could gain invaluable on-the-job experience in a fully operational front-line unit.

Maintenance at Sea

A Lynx detachment at sea was typically supported by a team of ten maintainers: five airframe/engine engineers, three avionics specialists and two armourers. One of these was team leader, while another one was dual-qualified as a rescue diver. According to Lieutenant de Vaisseau (Lieutenant) Jean-André (name withheld on demand), Flottille 34F Senior Engineering Officer:

> A team of ten maintainers was standard within the Lynx community. Its size had been determined a long time ago by the Lynx availability requirement at sea, the need for allocated assets to carry out the forecasted maintenance and the ship's capability in terms of accommodation and catering facilities. In order to sustain the Lynx at sea over long periods, the team could be divided into two watches, each composed of four men: two airframe/engine engineers, one avionics specialist and one armourer, although this could vary from one detachment to another. The team leader and the rescue diver were not included into the watches, as they also performed other key duties. Occasionally, a young engineer could join the team to reinforce it and build up experience at sea. It should be noted that when a frigate deployed with two Lynxes, the number of maintainers stayed at ten. This meant that when one rotorcraft was airborne, maintenance was carried out on the second one. The detachment was

Engineers at work on a Lynx on board *Dupleix*, off Libya, in 2011.

then composed of ten engineers, two Lynxes and two full crews, each with two pilots and one sonar operator/loadmaster.

At sea, each technician had to be multi-skilled, and Lynx detachment personnel had to support and help each other. 'For example, an avionics specialist could assist airframe/engine specialists changing a turbine or a rotor blade,' insists Maître Principal (Chief Petty Officer) Cédric, Flottille 34F deputy engineering officer. He continues:

> An avionics mechanic could carry out simple tasks usually performed by an airframe/propulsion technician and take full responsibility for it, albeit the control would still be the responsibility of a dedicated, fully qualified licence holder. This is the reason why we always sent to sea experienced maintainers who could perform a wide range of tasks by themselves. They must also be able to integrate smoothly into the frigate environment and to closely cooperate with the warship's crew members. Working atmosphere was usually excellent, and you could feel the camaraderie and team spirit.

The maintenance programme was exactly the same whether the Lynx was at sea or at its Lanvéoc-Poulmic home base because the base is so close to the Atlantic Ocean that it is considered a salty, corrosive environment. CPO Cédric continues:

> Lynx maintenance was optimised from the very beginning for naval operations at sea. At sea, the lack of space and the ship movements had to be taken into account as they made things more challenging. For example, rolling and pitching motions could be a problem when a Lynx had to be jacked up for interventions on the landing gear. The answer was to plan well in advance, to make sure that heavy maintenance would be undertaken in a safe and efficient manner. Sometimes, we had to conduct maintenance when the ship was pier-side, during a scheduled port visit.

Full Spare Allocation

On board a frigate, space is at a premium, and spares and support equipment had to be carefully stored using all available storage areas in the vicinity of the hangar. CPO Cédric recalls:

> A large amount of equipment and supplies was loaded on board the surface combatant prior to sailing. It included a comprehensive range of tools, a main rotor blade, a tail rotor blade, a spare Gem turbine, tyres, various black boxes, a rotor brake, a pylon for the MU90 torpedo, a sonar, a winch, decoy dispensers, a pintle-mount for an ANF1 7.62mm machine gun, an articulated arm for a McMillan, Barret or PGM Hécate 12.7mm sniper rifle, a Chlio FLIR turret, an abseiling/fast roping kit, a vertrep kit, desert/arctic aircrew survival equipment and various supplies such as lubricants, glues and aeronautical sealants.

A robust logistics network was required to support a Lynx detachment at sea. Lieutenant Jean-André explains:

> When detachments deployed overseas for an extended period of time, we had to schedule a number of resupplies at regular intervals to ensure that their Lynx remained serviceable. It was pure operational logistics and, depending on the level of priority and of urgency, we could dispatch spares using a military aircraft, a commercial flight or by contracting Fedex, DHL or any other multinational delivery services. Should the need arise, we could even air drop from an Atlantique 2 maritime patrol aircraft a small container that would be retrieved by a rigid-hull inflatable craft launched from the frigate.

Above left: A Flottille 34F Lynx undergoing maintenance in the squadron's hangar, in BAN Lanvéoc-Poulmic. Noteworthy are all the open access panels. Accessibility to components was considered to be good on the Lynx.

Above right: The open cowling reveals a Rolls-Royce Gem turbine. Over the years, this engine has proved extremely powerful and remarkably reliable.

Lynx s/n 802 sits on the Flottille 34F ramp between two sorties in June 2019.

Two engineers working side by side to replace a seal on the undercarriage of the Lynx. The main wheels have already been towed out for operations from the deck of a frigate, allowing the helicopter to perform a spot turn into wind while the harpoon remains engaged into the deck steel grid. A job in a Lynx squadron was a highly sought-after posting within the French Navy aviation engineering community.

Above: In French service, the Lynx has proved to be a rugged, reliable and dependable aircraft. This Flottille 34F helicopter performs a high-speed pass.

Left: Like all metallic aircraft at sea, the Lynx had to be protected from corrosion, and engineers spent a lot of time and effort fighting the assaults and effects of salt water.

Lynx s/n 273 with its rotor and tail boom folded inside the *Latouche-Tréville*'s helicopter hangar. The flag hanging on the left is the flag of Brittany, the region where Flottille 34F is stationed.

Great care had to be taken when stowing a Lynx on board a ship. The rotor blades were extremely fragile and needed to be folded to reduce the helicopter's footprint. Here, a maintainer locks the tail rotor prior to folding the tail boom.

Right: Maintainers are preparing to manually fold this Lynx's main rotor aboard *Latouche-Tréville* in July 2020.

Below: Lynxes are a tight fit inside F70-class frigates' hangars.

Maintainers are busy preparing two Lynxes parked inside *Latouche-Tréville*'s dual hangar. This photo clearly shows why the much bigger NH90 Caïman cannot operate from F70-class ASW frigates.

Left: Close-up on the folded tails of the two Lynxes sitting side by side inside the *Latouche-Tréville* during the Lynx's last detachment at sea. The tail rotor transmission can clearly be seen.

Below left: For added endurance and range, the Lynx could be fitted with an additional internal fuel tank at the expense of some volume in the already cramped main cabin.

Below right: Lynx sitting on a slope, atop a cliff on the Brittany coastline. It was then flown by Lieutenant Commander Christophe, Flottille 34F's operations officer.

Above: Navy personnel attach a long-line to a store during a sling-load training sortie.

Right: The atmosphere within Lynx detachments was generally very good, with an excellent working ambiance. Maintainers were all highly motivated and good humour prevailed.

Up to 2,425lbs (1,100kg) could be carried by a Lynx under sling, although a 1,102lbs (500kg) load was an operational maximum under normal circumstances.

Chapter 8

Final Lynx Deployment at Sea

The last Lynx deployment at sea was carried out on board *Latouche-Tréville* from 12 to 23 July 2020 during a mission dedicated to the escort of a FOST nuclear submarine and to the protection of classified trials by the DGA, the French defence armament and procurement agency.

After the withdrawal from service of the *La Motte Piquet*, in May 2020, the *Latouche-Tréville* had become the last surviving F70-class frigate and the last ship equipped with Lynx rotorcraft in France. 'For us, surface warfare specialists, the withdrawal of the Lynx is a highly emotional moment,' says Capitaine de Vaisseau (Captain) Patrick de Sevin, commanding officer of the *Latouche-Tréville*. He continues:

During the Cold War, the frigate/Lynx duo was among the most effective ASW tools anywhere in the world, and that could be said of both F67 *Tourville*-class and F70 *Georges Leygues*-class frigates. What is remarkable is that the duo remained fully operational for over 40 years. Under current plans, the *Latouche-Tréville* will stay in service for another two years, until 2022. For a surface combatant, losing her ASW helicopter is not a trivial concern, but, for future missions, we will be equipped with a Panther, a Dauphin or an Alouette III, even though none of these rotorcraft are fitted with a dunking sonar. Thankfully, the state-of-the-art *Aquitaine*-class multi-mission frigates and their NH90 Caïmans are already there to handle ASW missions.

Flottille 34F's last Lynx detachment personnel and *Latouche-Tréville* flight deck personnel pose on the last day of Lynx operation at sea, on 23 July 2020.

For the last mission, off Brittany, a single Lynx, s/n 273, had deployed on board, but it was intermittently reinforced by s/n 272 that shuttled back and forth between the frigate and its Lanvéoc-Poulmic home base. Lynx s/n 622 also landed on numerous occasions on the frigate during that week, including twice for photo flights with the author of this book on board. The Lynx was then flown by Capitaine de Corvette (Lieutenant Commander) Sébastien, the deputy commander of Flottille 34F.

Optimised for Operations at Sea

Commander Chaput took advantage of the deployment to the frigate to explain to us how the Lynx behaved in a maritime environment:

> The Lynx was conceived from the very beginning for operations at sea. Westland engineers carefully designed the airframe and its propulsion system to give aircrews excellent agility and power margin in the landing circuit, even when operating in the extreme corners of the flight envelope, in high winds. Operations could be undertaken up to sea state 6, but, in worse sea states, things rapidly became more complicated and we had to take a closer look at performance and deck-landing diagrams. Our operational wind limit was 50 knots, but we had to keep in mind that, if the Lynx easily accepted wind coming from the left because the tail rotor was on the left side of the tail boom, it was far more tricky to fly in strong winds from the right. *Latouche-Tréville*, like all F70 *Georges Leygues*-class frigates, is perfectly adapted to the sea conditions in the Atlantic: she offers excellent sea-keeping qualities and proves very stable. As a result, it was always a pleasure to come and land on her platform.

Flown by Commander Chaput, Lynx s/n 272 comes in to land on the *Latouche-Tréville*'s helicopter platform.

The platform is bigger on the F70s than on the earlier F67s as Commander Chaput explains:

> On the now-retired *Tourville*-class vessels, the steel grid into which the harpoon locked to secure the helicopter on the deck was much closer to the hangar door than on later *Georges Leygues*-class frigates. We had a feeling of close proximity to the ship superstructures, the same our Panther colleagues have when landing on the *Jean Bart* air-defence destroyer, a ship with a notoriously small platform by modern standards. We had to be careful and extremely precise when hovering and coming in to land.

Very Good Handling Characteristics

The Lynx's agility and power were obvious advantages for operations at sea. Commander Chaput continues:

> In a Panther or a Dauphin, you will very soon reach the aircraft's power limits, especially at heavy weights in hot weather, typically in the Persian Gulf in the summer. On these types, when close to the power limit, we are committed to landing and have to fly a straight in approach, without hovering over the platform, flying all the way down until landing without any pause. On the Lynx, we basically never reached the power limit, even in very hot weather. This meant we could come in and stop the descent to have a closer look at the platform and ship movements and behaviour. We had more options as it gave us a safety margin and the opportunity to push the limit and go a little bit further in terms of weather conditions as we could wait above the deck for the ideal moment to come in to land.

The Lynx's semi-rigid rotor head is the key technology that helped significantly increase the aircraft's agility. Capitaine de Corvette (Lieutenant Commander) Christophe, Flottille 34F's operations officer, adds:

> The semi-rigid rotor head allowed us to fly the aircraft extremely accurately around the boat. This proved to be a crucial advantage which boosted our confidence in adverse conditions. The Lynx instantly reacted to control inputs, with zero latency. It was very pleasant and pretty comfortable, but young pilots had sometimes a tendency to over control and we advised them to fly using the trim control only in some critical phases of the flight, particularly during the final stage of the descent, when stabilising over the platform.

2,000th Night Deck Landing

During the last deployment at sea of the Lynx, Commander Chaput passed a significant personal milestone when he logged his 2,000th night deck landing, a major accomplishment. Commander Chaput

Commander Chaput flying the Lynx in the course of the sortie during which he logged his 2,000th deck landing at night, an incredible achievement. The instrument panel is dated by modern standards.

is the French Navy's active duty rotary pilot still serving with the highest number of flying hours to his credit (7,000 flight hours, including 2,200 in the Lynx). The milestone was reached during a training mission on 22 July 2020 from frigate *Latouche-Tréville* sailing off Brittany. During a series of touch-and-goes, his personal tally reached 2,004 night-time deck landings. Eventually, this tally was expanded to 2,005 landings after his Lynx was scrambled later in the night for a live medical evacuation towards a hospital in Brest. It should be noted that Commander Chaput has also performed 3,757 deck landings in daylight, bringing his daylight/night-time tally to an amazing 5,762 deck landings! 'I am not totally sure, but I might be the pilot credited with the highest number of deck landings in the whole history of the Marine Nationale,' he said. 'I have always been lucky during my career, obtaining posting or deploying during missions that would help me log large amount of flying hours.'

Last Landing at Sea

The French Lynx performed its last ever landing at sea on board *Latouche-Tréville* on 23 July 2020 off Brest during a series of touch-and-goes. The crew was composed of Commander Chaput (captain), Lieutenant Pierre (co-pilot) and Maître Guirec (petty officer crewman/loadmaster). The Lynx immediately took off again, with the crew of *Latouche-Tréville* manning the rails. Commander Chaput flew alongside the vessel, rocking the Lynx from left to right and from right to left for a final goodbye before heading back towards Lanvéoc-Poulmic. Obviously very moved, Commander Chaput said after landing that 'This was the last landing at sea for a French Lynx, a major historical event for the Marine Nationale. From a more personal point of view, this was probably my last landing ever at sea as I am now posted to Dax to fly the EC120 Colibri at the joint helicopter training school where I will not have the opportunity to deploy at sea anymore.'

Replaced by the NH90 Caïman

The very last flight of the Lynx was carried out on 4 September 2020 at Lanvéoc-Poulmic, after 42 years of sterling service. With this proven and dependable helicopter, the Marine Nationale has developed and expanded advanced operational ASW tactics. However, it has to be admitted that the type was fast ageing and that its sensors offered only limited detection capabilities compared to those of its successor, the state-of-the-art NH90 Caïman. The first of 27 Caïmans was delivered to the French Navy in April 2010. The order was split into 14 of the NHC (C for Combat) variant to supplant the Lynx, and 13 of the

The NH90 Caïman has now replaced the Lynx as the French Navy's main rotary ASW asset. This aircraft belongs to Flottille 31F at Hyères.

ramp-equipped NHS (S for Soutien, or Support) version. This split is in fact a bit misleading as the two variants can be equipped with the Thales Underwater System FLASH (Folding Light Acoustic System for Helicopters) sonar and with MU90 torpedoes for ASW missions. The type reached IOC (Initial Operational Capability) in December 2011 with Flottille 33F, at Lanvéoc-Poulmic, and FOC (Full Operational Capability) two years later. Since then, the NH90 Caïman has proved to be a very effective ASW tool, its FLASH sonar offering at least five times the detection range of the Lynx's DUAV-4 sonar. As expected from a much bigger aircraft, the Caïman also offers a considerably longer endurance and a better payload capability, with up to two MU90s being routinely carried instead of one for the Lynx.

New Future for Flottille 34F

At the time of writing, it was expected that Flottille 34F would be recreated in January 2021 to operate the four leased Dauphins used for helicopter pilot/rear crew training and the last few Alouette IIIs still in service for various support duties. It would thus take over Escadrille 22S' role, with 22S' number plate then disappearing. Once the last Alouette IIIs are withdrawn circa 2023, the new Flottille 34F will operate a fleet of around ten leased Dauphins as, to fill the gap between the Lynx and the Alouette III on the one hand and the future H160M Guépard on the other hand, the French Navy has committed to fielding an interim fleet of helicopters composed of 16 leased Dauphins (to be shared between Flottilles 34F and 35F) and four H160 helicopters to be operated by Flottille 32F. These civilian H160s will supplant the NH90s used for SAR missions from Cherbourg and Lanvéoc-Poulmic, allowing the Caïmans to focus on combat missions. They will also give Marine Nationale aircrews and maintainers the opportunity to gain experience on the type until the delivery of the first armed H160M to the French Navy in 2028. Flottille 34F leased Dauphin N3/N3+ helicopters will all be in service by 2023 and will be cleared to operate from frigates, and, as such, some of them could find themselves embarking on board the *Latouche-Tréville*. Eventually, Flottille 34F will convert to the H160M Guépard in the late 2020s/early 2030s.

This member of the flight deck crew secures a restraining chain to the Lynx on board *Latouche-Tréville*. During the last deployment, activity at sea was fairly intensive.

Above left: Lynx s/n 622 flown by Lieutenant Commander Sébastien, the deputy commander of Flottille 34F. Noteworthy is the grey, radar-compatible nosecone.

Above right: Lynx s/n 273 was photographed in June 2020, three weeks before the last deployment at sea. At the time, it had been fitted with a black radome for a photo flight.

Right: The Lynx has already been secured to the flight deck. Aircrews were all wearing face masks, in accordance with the French Navy's Covid regulation at the time.

Below: A very last salute by Commander Chaput. The photo ship Lynx was flown by Lieutenant Commander Sébastien.

Above: A fireman in heavy-duty gear keeps an eye on the Lynx, ready to intervene at the first sight of smoke or a flame.

Left: Folding the main rotor blades of a Lynx is done manually. It is a labour-intensive task that requires at least four engineers. Here, the weather is perfect. Imagine the same task performed in the cold, in the rain, on a pitching deck, at night and in the middle of winter, north of the Arctic Circle.

Commander Chaput has just disembarked from his Lynx after a trip to Lanvéoc-Poulmic and back. The maintainer in the foreground is positioning a stepladder that will be used to help fold the main rotor blades.

Above left: A Lynx waiting for its crew before a night training sortie on board *Latouche-Tréville*. A large percentage of the flying hours allocated to Flottille 34F were flown at night.

Above right: Frigate *Latouche-Tréville* seen from astern. The author tweaked his camera to obtain that photo. In fact, the night was much darker.

Right: The yellow shirt guiding the Lynx can just be seen at the far corner of the hangar. The weather was excellent that day. Landing on board *Latouche-Tréville* was no big deal for such an experienced naval aviator as Commander Chaput.

Below: François Chaput at the end of his record-breaking sortie. Even though it is the end of July, he is wearing an immersion suit... and a combat dagger fastened to his right leg. In an emergency, after a crash, a ditching or a fire, it could be used to cut him free from his harness.

Commander Chaput overflies ASW frigate *Latouche-Tréville* in gorgeous weather. The crew is manning the rail in their best uniform to honour the end of the Lynx.

The last ever landing of a French Lynx on board a Marine Nationale surface combatant.

Commander Chaput demonstrates the Lynx's agility, pushing the nose over into a dive.

Left: Commander Chaput's Lynx look insignificant compared to the sheer size of the *Latouche-Tréville*.

Below: The last day of a very efficient ASW duo. The Lynx will undoubtedly be missed.

Above: The very last take-off of a French Lynx from a French warship, on 23 July 2020, at the mouth of the Brest Bay. A very moving moment for all who witnessed the event.

Right: Compared to the Lynx, the Caïman is a much bigger aircraft that offers a higher payload and a longer range.

The NH90 Caïman is fitted with the latest generation of sensors, including an active low-frequency dunking sonar, a multifunction radar and an ESM suite.

Above left: For pilots transitioning from the Lynx to the Caïman, the NH90's avionics are undoubtedly a major step forward.

Above right: The sensor operator/sonar operator in the back has at his disposal an advanced man-machine interface and the latest generation of displays.

Above left: Full-scale mock-up of the future H160M Guépard exposed at Lanvéoc-Poulmic in September 2020.

Above right: Close-up on the Sea Venom/ANL missile and on the door-mounted articulated arm fitted with a 12.7mm precision rifle.